What's Inside Earth?

Jane Kelly Kosek

The Rosen Publishing Group's
PowerKids Press™
New York

For Kaitlin—may you enjoy the wonders of Earth.
Special thanks to Christopher H. Starr at the Pacific Sierra Research Corporation.

Published in 1999 by The Rosen Publishing Group, Inc.
29 East 21st Street, New York, NY 10010

First Edition

Book Design: Kim Sonsky

Photo Credits: Cover, back cover, title page, contents page, pp. 6, 21 © Digital Vision Ltd., p. 5 © FPG International/Telegraph Colour Library, p. 6 girl by Seth Dinnerman, p. 13 © FPG International/ Philip Wallick, p. 14 © FPG International/Robert Reiff, p. 15 © International Stock/Michele and Tom Grimm, p. 16 © FPG International/Howard Davis, p. 17 © FPG International/David Sacks, p. 17 © International/Al Clayton, p. 18 © PhotoDisc.

Kosek, Jane Kelly.
What's inside Earth? / by Jane Kelly Kosek. — 1st ed.
p. cm. — (What's inside?)
Includes index.
Summary: Discusses the layers, water, atmosphere, and future of Earth.
ISBN 0-8239-5277-0
1. Earth—Internal structure—Juvenile literature. [1. Earth.]
I. Title. II. Series: Kosek, Jane Kelly. What's inside?
QE509. K67 1998
551.1—dc21 97-52820
CIP
AC

Manufactured in the United States of America

Contents

Earth's Beginnings

Earth is the **planet** (PLAN-et) on which we live. It is a part of a **solar system** (SOH-ler SIS-tem). Our solar system is made up of the sun, nine planets (including Earth), moons, and other space objects. Earth is the third planet from the sun. It **orbits** (OR-bits), or travels around, the sun. It takes Earth 365 days, or one year, to travel around the entire sun.

Earth began forming about 4.5 billion years ago, after the sun first formed. After the sun was born, leftover gas and dust began to orbit it. Some of this gas and dust slowly came together to form Earth. Earth is special because it is the only planet that we know to have life, including you!

We have nine planets in our solar system: Mercury, Venus, Earth, Mars, Jupiter, Saturn, Uranus, Neptune, and Pluto.

Earth's Layers

Earth has many layers. Some layers are solid, such as the crust, or liquid, such as the ocean, or gas, such as the sky. When Earth was forming, **gravity** (GRAV-ih-tee) pulled heavier things, like solids and liquids, toward the middle of Earth. This is why Earth has solid and liquid layers first and then a layer of gases.

Earth's first layer, starting at its center, is called the **core** (KOR). Around the core lies the **mantle** (MAN-tul). Around the mantle is an outer shell containing the crust. The crust is what we live on and is where the ocean lies. The last layer of Earth is the **atmosphere** (AT-mus-feer). The atmosphere is what we breathe every day.

◀ You may not know it, but you walk on top of many layers of the earth.

The Core

Earth's core, or center, is about 2,000 miles below Earth's **surface** (SER-fiss). That's pretty far. It would take you about 36 hours to drive there if you could.

The core has two layers: the inner core and the outer core. The inner core is solid, and the outer core is hot liquid. The outer core is about 20 times larger than the inner core. Both layers are made mostly of iron. The **temperature** (TEMP-rah-cher) of Earth's inner core is about 7,500 degrees. We don't feel that heat, though, because Earth's thick layers keep most of the heat from reaching the surface.

The solid inner core at the center of Earth is surrounded by a liquid outer core.

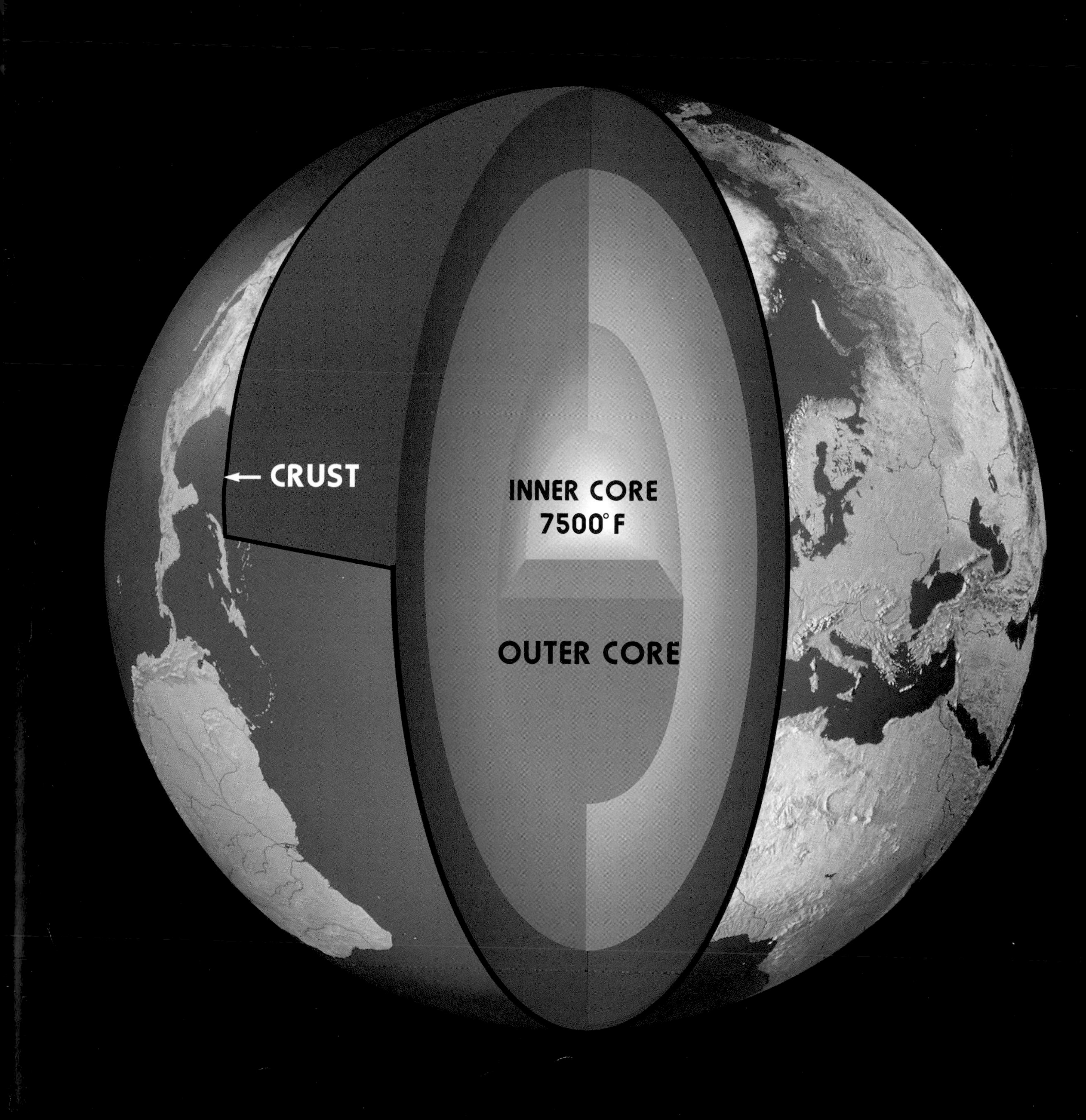
CRUST
INNER CORE
7500°F
OUTER CORE

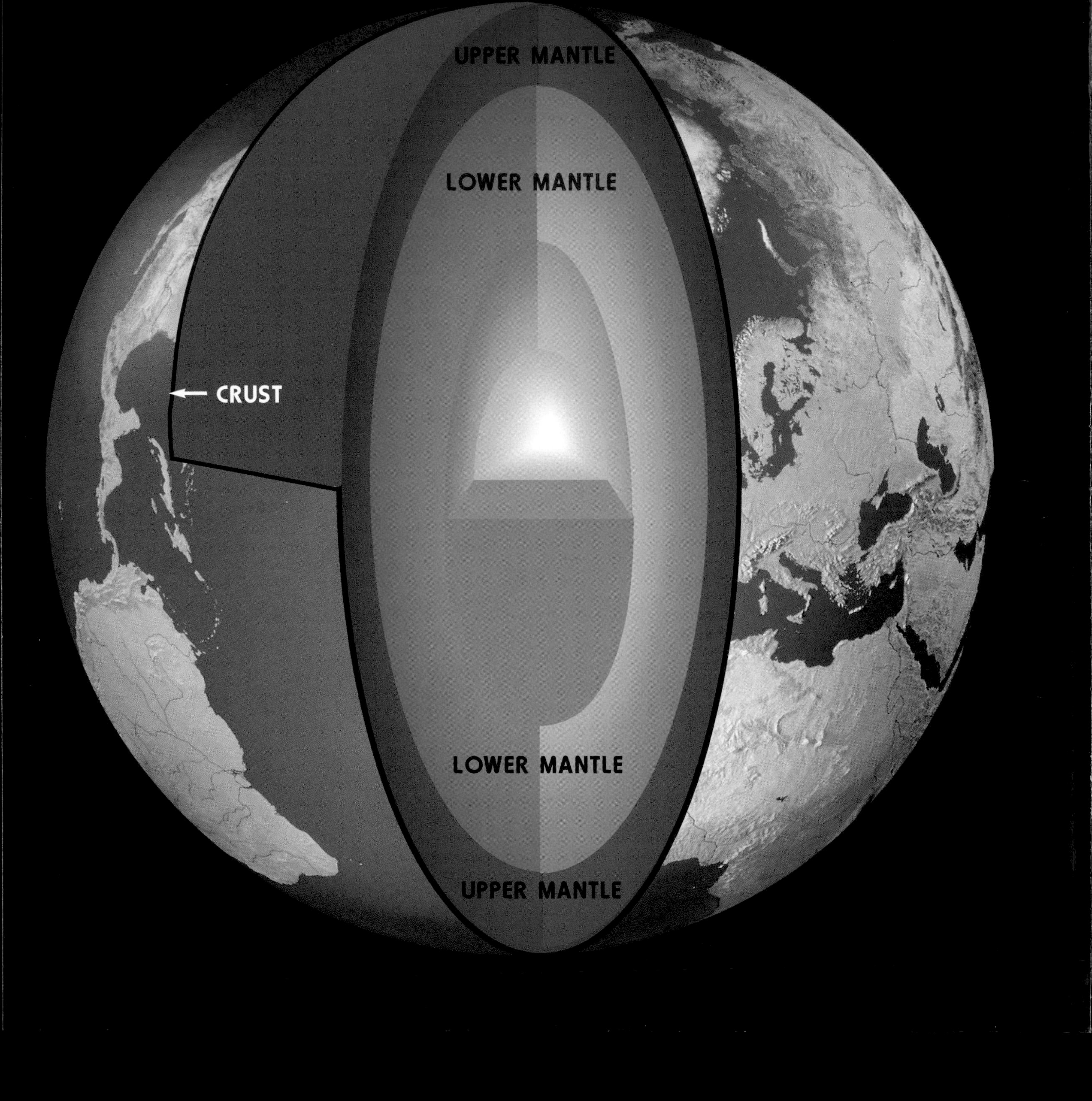
UPPER MANTLE
LOWER MANTLE
CRUST
LOWER MANTLE
UPPER MANTLE

The Mantle

The mantle rests on top of the outer core. It is located anywhere from 6 to 40 miles under Earth's surface and is about 1,800 miles thick. The heat from the core makes the mantle very hot. This layer has two areas: the lower mantle and the upper mantle.

The lower mantle is solid rock. It has the greatest **mass** (MAS) of any layer in Earth. The upper mantle is also solid rock with a thin layer that is part liquid. This liquid layer moves slowly. This causes Earth's surface to change, or move.

A thin liquid layer in the mantle causes the Earth's surface to move. But it moves much too slowly for us to feel it.

Earth's Moving Surface

Earth's surface is always moving. This surface is made of separate pieces called plates. They rest on top of the moving liquid layer in the upper mantle.

Eight major plates make up the entire surface of Earth. Each plate is made of a piece of the upper mantle and the crust, and each is named for the area it is in. For example, the United States, which is in North America, is part of the North American Plate.

Because plates move, their edges can run into each other. They can slide next to each other in opposite directions, or even spread away from each other. This movement doesn't happen very quickly, though. A plate will only move about two to four inches per year.

The San Andreas fault in California is where two plates of Earth's crust meet.

The Crust

We, along with billions and billions of other living things, live on Earth's crust. Earth's crust is made of many different kinds of rocks. The part of the crust that we live on is called the continental crust. It is 20 to 40 miles thick. The part of the crust that the ocean rests on is called the oceanic crust. It is about 6 miles thick. Most of the heat from inside Earth does not reach Earth's surface. Some of the heat is released through the crust by a **volcano** (vol-KAY-noh). When a volcano erupts, **magma** (MAG-muh), or melted rock, rises through the crust and is released as **lava** (LAH-vuh). This lava then cools and forms new crust.

Smoke and steam are also released when volcanoes such as Kilauea in Hawaii erupt.

Looking Inside the Crust

We use many things found inside Earth's crust. Oil, coal, and natural gas come from the crust. We use oil to help cars run. Coal and natural gas may be used to make heat. These three things are thought to be the **remains** (re-MAYNZ) of living things from long, long ago. Scientists believe that coal comes from the remains of ancient forests. Oil and natural gas could be from the remains of ancient life in the ocean. Gemstones, including diamonds, rubies, and sapphires, are also found in Earth's crust. These stones are often worn by people as jewelry.

Diamonds are created when the pressure inside Earth pushes on coal over thousands of years.

Water

Did you know that 71 percent of Earth's crust is covered by ocean water? The other planets nearest to us do not even have water on their surfaces. But there are two kinds of water on Earth. The water in our ocean is called salt water because it contains a large amount of salt. Rivers, streams, ice, and most lakes on Earth's surface contain fresh water. We need fresh water to live. Rain is the **source** (SORSS) of our fresh water. Rain is formed when the heat from the sun causes water on Earth to **evaporate** (ee-VAP-or-ayt). This water later falls from the sky as rain. Many living things, such as fish, land animals, and plants, rely on both kinds of water to live.

Many different kinds of animals, including fish, live in the ocean.

Atmosphere

When you look at the sky, you are looking through our atmosphere. The atmosphere is a layer of gases that surrounds Earth. The gas that our bodies need to breathe is called **oxygen** (AHK-sih-jen). It makes up part of our atmosphere. We breathe in this gas every day.

The sun shines down on our atmosphere, giving us heat, light, and energy. The atmosphere protects us from the harmful rays given off by the sun. Clouds are also part of our atmosphere. Rain that forms from the evaporated water of Earth's surface falls from clouds.

The state of the atmosphere can change every day. It may be sunny one day and rainy the next.

The Future of Earth

You have learned what goes on inside our planet Earth. Layers are moving and changing under us. It can be exciting to know about the things going on under Earth's crust. But we should also remember how important our planet is to us.

We live on a very small part of Earth. But what we do affects everything on the planet. Some people have hurt Earth by cutting down trees in our rain forests. Others have polluted the air and the water.

Everyone needs to **respect** (re-SPEKT) our planet. You can help by learning to **recycle** (re-SY-kul) and not being a litterbug. The future of Earth is up to us.

Web Sites:

To learn more about Earth, check out these Web sites:
http://bang.lanl.gov/solarsys/earthint.htm
http://seds.lpl.arizona.edu/nineplanets/nineplanets/earth.html

Glossary

atmosphere (AT-mus-feer) A layer of gases around Earth.
core (KOR) The center of Earth.
evaporate (ee-VAP-or-ayt) To change from a liquid to a gas.
gravity (GRAV-ih-tee) A force between two objects that causes them to be attracted to each other.
lava (LAH-vuh) What magma is called when it reaches Earth's surface.
magma (MAG-muh) A hot, liquid rock inside the mantle.
mantle (MAN-tul) A layer of Earth between the core and the crust.
mass (MAS) The amount of material in something.
orbit (OR-bit) When one thing circles another.
oxygen (AHK-sih-jen) A colorless gas that forms part of the air we breathe.
planet (PLAN-et) A large object that orbits the sun.
recycle (re-SY-kul) To use again.
remains (re-MAYNZ) What is left after a plant or an animal has died.
respect (re-SPEKT) To think highly of something or someone.
solar system (SOH-ler SIS-tem) The system made up of our sun, the nine planets, moons, and other space objects.
source (SORSS) Where something comes from.
surface (SER-fiss) The top or outside of something.
temperature (TEMP-rah-chur) How hot or cold something is.
volcano (vol-KAY-noh) An opening in Earth's crust through which magma is sometimes forced out.

Index